What Kind of Clouds?

水无处不在 那是什么云

[美] 纳迪亚·希金斯 著

[美] 莎拉·英芬特 绘

[美] 艾瑞克·科斯金恩 作曲

沈可君 译

中国水利水电出版社

www.waterpub.com.cn

·北京·

项目策划：徐丽娟
责任编辑：栾 峰 方 斯
特约编辑：李渝汶
联系方式：luanfeng@mwr.gov.cn 010-68545978

书 名 水无处不在 那是什么云
SHUI WUCHUBUZAI NA SHI SHENME YUN

作 者 [美]纳迪亚·希金斯 著 [美]莎拉·英芬特 绘

[美]艾瑞克·科斯金恩 作曲 沈可君 译

出版发行 中国水利水电出版社

（北京市海淀区玉渊潭南路1号D座 100038）

网 址：www.waterpub.com.cn

E-mail：sales@mwr.gov.cn

电 话：（010）68367658（营销中心）

经 售 北京科水图书销售中心（零售）

电 话：（010）88383994、63202643、68545874

全国各地新华书店和相关出版物销售网点

排 版 陆 云

印 刷 北京尚唐印刷包装有限公司

规 格 285mm×210mm 16开本 6印张（总） 80千字（总）

版 次 2022年1月第1版 2022年1月第1次印刷

总 定 价 148.00元（全4册）

凡购买我社图书，如有缺页、倒页、脱页的，本社营销中心负责调换

版权所有·侵权必究

图书在版编目（CIP）数据

水无处不在. 那是什么云：汉英对照 /（美）纳迪亚·希金斯著；沈可君译. -- 北京：中国水利水电出版社，2022.1
书名原文：Water All Around Us
ISBN 978-7-5226-0105-2

Ⅰ. ①水… Ⅱ. ①纳… ②沈… Ⅲ. ①水—儿童读物—汉、英 Ⅳ. ① P33-49

中国版本图书馆 CIP 数据核字（2021）第 210674 号

北京市版权局著作权合同登记号：图字 01-2021-5364

亲子学习诀窍

为什么和孩子一起阅读、唱歌这么重要？

每天和孩子一起阅读，可以让孩子的学习更有成效。音乐和歌谣，有着变化丰富的韵律，对孩子来说充满乐趣，也对孩子生活认知和语言学习大有助益。音乐可以非常好地把乐感和阅读能力锻炼有机结合，唱歌可以帮助孩子积累词汇和提高语言能力。而且，在阅读的同时欣赏音乐也是增进亲子感情的好方式。

记住：要每天一起阅读、唱歌哦！

绘本使用指导

1. 唱和读的同时找出每页中的同韵单词，再想想有没有其他同韵单词。
2. 记住简单的押韵词，并且唱出来。这可以培养孩子的综合技能以及英语阅读能力。
3. 最后一页的"读书活动指导"可以帮助家长更好地为孩子讲故事。
4. 跟孩子一起听歌的时候可以把歌词读给孩子听。想一想，音符和歌词里的单词有什么联系？
5. 在路上，在家中，随时都可以唱一唱。扫描每本书的二维码可以听到音乐哦。

每天陪孩子读书，是给孩子最好的陪伴。
祝你们读得快乐，唱得开心！

扫我听音乐

There are many kinds of clouds. Some are white and fluffy. Others are thin and gray.

Clouds can help you to tell what the weather will be like. But no matter what kind of

clouds are floating up in the sky, all clouds are made of water droplets.

Turn the page to learn all about the different kinds of clouds. Remember to sing along!

天空中有各种各样的云，有蓬松的白云，也有薄薄的乌云。

我们可以通过云判断天气。

但无论天空中飘着什么云，

它们都是由水滴构成的。

请翻到下一页，我们一起来认识云的种类。

跟着音乐一起唱吧！

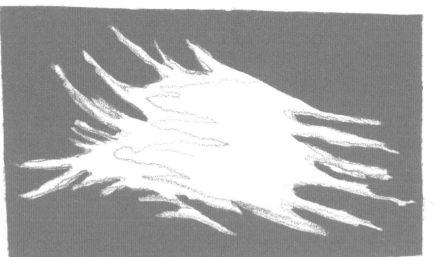

Clouds are water. It's that simple.
Vapor floats up in the sky.
Tiny drops and icy crystals
form the sky shapes drifting by.

云就是水，简单道理。
水汽蒸腾，飞上天际。
小小水滴，点点冰晶，
形状各异，空中飘起。

Are they gray or white, fluffy or thin?
What kind of clouds will this day bring?

那云是灰，是白？
是蓬松，是薄轻？
今日天气，有什么云？

A stratus cloud!

那是层云！

Like a blanket, this cloud lies
low and thin across the sky.
Watch it drizzle, oh, so gray.
Settle in for a cozy day.

薄薄的云，低低伸展，铺在天空，好似毛毯。
毛毛细雨，天色灰灰，舒适惬意，一天懒懒。

9

Clouds are water. It's that simple.

Vapor floats up in the sky.

Tiny drops and icy crystals

form the sky shapes drifting by.

云就是水，简单道理。

水汽蒸腾，飞上天际。

小小水滴，点点冰晶，

形状各异，空中飘起。

Are they gray or white, fluffy or thin?

What kind of clouds will this day bring?

那云是灰，是白？

是蓬松，是薄轻？

今日天气，有什么云？

A cirrus cloud!

那是卷云！

So high above and icy white,
it sparkles in the bright sunlight.
Thin and wispy as a feather,
cirrus promises fair weather.

云丝雪白，高高飘起。太阳照耀，晶亮莹莹。
轻盈透薄，好似飞羽。卷云一现，明日天晴。

Clouds are water. It's that simple.

Vapor floats up in the sky.

Tiny drops and icy crystals

form the sky shapes drifting by.

云就是水，简单道理。水汽蒸腾，飞上天际。

小小水滴，点点冰晶，形状各异，空中飘起。

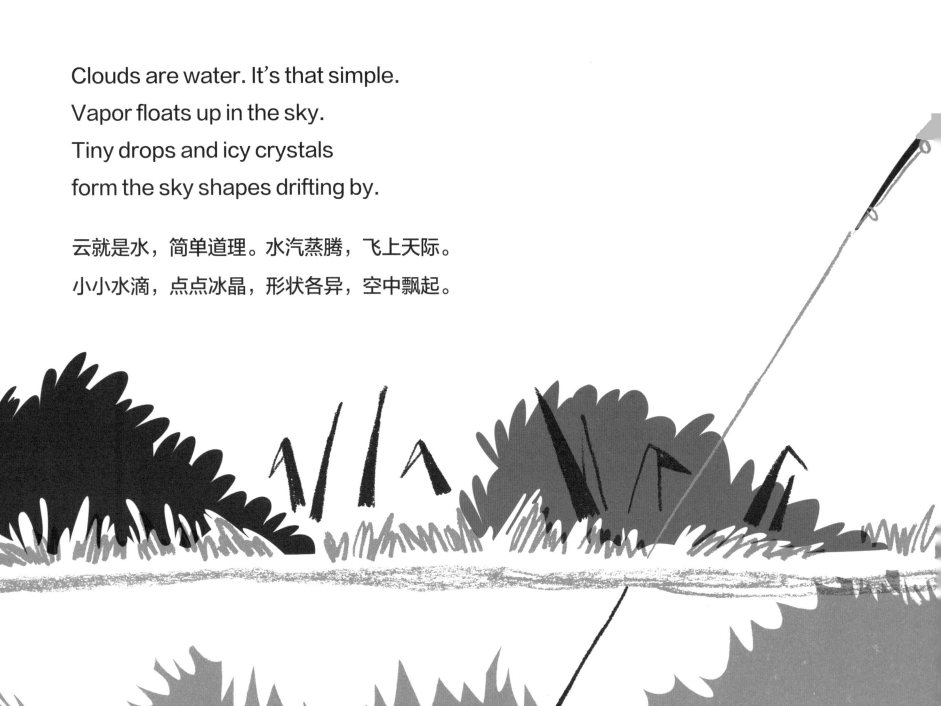

Are they gray or white, fluffy or thin?

What kind of clouds will this day bring?

那云是灰，是白？

是蓬松，是薄轻？

今日天气，有什么云？

A cumulus cloud!

那是积云！

White and puffy, it drifts through
a sky so perfectly blue,
shifting fast from shape to shape—
a cat, a dragon, now an ape!

朵朵白云，软软轻轻。
映衬蓝天，自在游弋。
千姿百态，形状不定。
变猫，变龙，再变猩猩！

Clouds are water. It's that simple.
Vapor floats up in the sky.
Tiny drops and icy crystals
form the sky shapes drifting by.

云就是水，简单道理。水汽蒸腾，飞上天际。
小小水滴，点点冰晶，形状各异，空中飘起。

Are they gray or white, fluffy or thin?

What kind of clouds will this day bring?

那云是灰，是白？是蓬松，是薄轻？

今日天气，有什么云？

A cumulonimbus cloud!

那是雨云！

Towering tall and flat on top,

this storm cloud is ready to pop.

Flash! Crack! Growl! Ka-BOOM!

Thunder and lightning shake the room.

Are they gray or white, fluffy or thin?

What kind of clouds will this day bring?

平悬天空，高耸穹隆。云中暴雨，随时倾下。

忽闪！咔嚓！轰隆作响！雷鸣电闪，撼动万家。

那云是灰，是白？是蓬松，是薄轻？

今日天气，有什么云？

21

SONG LYRICS 歌词
What Kind of Clouds?

Clouds are water. It's that simple.
Vapor floats up in the sky.
Tiny drops and icy crystals
form the sky shapes drifting by.

Are they gray or white, fluffy or thin?
What kind of clouds will this day bring?

A stratus cloud!
Like a blanket, this cloud lies
low and thin across the sky.
Watch it drizzle, oh, so gray.
Settle in for a cozy day.

Clouds are water. It's that simple.
Vapor floats up in the sky.
Tiny drops and icy crystals
form the sky shapes drifting by.

Are they gray or white, fluffy or thin?
What kind of clouds will this day bring?

A cirrus cloud!
So high above and icy white,
it sparkles in the bright sunlight.
Thin and wispy as a feather,
cirrus promises fair weather.

Clouds are water. It's that simple.
Vapor floats up in the sky.

Tiny drops and icy crystals
form the sky shapes drifting by.

Are they gray or white, fluffy or thin?
What kind of clouds will this day bring?

A cumulus cloud!
White and puffy, it drifts through
a sky so perfectly blue,
shifting fast from shape to shape—
a cat, a dragon, now an ape!

Clouds are water. It's that simple.
Vapor floats up in the sky.
Tiny drops and icy crystals
form the sky shapes drifting by.

Are they gray or white, fluffy or thin?
What kind of clouds will this day bring?

A cumulonimbus cloud!
Towering tall and flat on top,
this storm cloud is ready to pop.
Flash! Crack! Growl! Ka-BOOM!
Thunder and lightning shake the room.

Are they gray or white, fluffy or thin?
What kind of clouds will this day bring?

What Kind of Clouds?

Chorus

Clouds are wa - ter. It's that sim - ple. Va-por floats up in the sky. Ti-ny drops and i - cy crys - tals form the sky shapes drift-ing by. Are they gray or white, fluff - y or thin? What kind of clouds will this day bring?

Verse

1. A stra - tus cloud! Like a blan - ket, this cloud lies low and thin a - cross the sky. Watch it driz - zle, oh, so gray. Set - tle in for a co - zy day.

Chorus

Verse 2
A cirrus cloud!
So high above and icy white,
it sparkles in the bright sunlight.
Thin and wispy as a feather,
cirrus promises fair weather.

Chorus

Verse 3
A cumulus cloud!
White and puffy, it drifts through
a sky so perfectly blue,
shifting fast from shape to shape—
a cat, a dragon, now an ape!

Chorus

Verse 4
A cumulonimbus cloud!
Towering tall and flat on top,
this storm cloud is ready to pop.
Flash! Crack! Growl! Ka-BOOM!
Thunder and lightning shake the room.

Outro

Are they gray or white, fluff - y or thin? What kind of clouds will this day bring?

GLOSSARY　　词汇表

crystals—glass-like substances with many see-through sides

冰晶——像玻璃一样的多面透明体

drifting—slowly moving

漂浮——缓慢移动

drizzle—rain that falls lightly in very small drops

细雨——慢慢落下的细小雨滴

puffy—rounded and airy

蓬松——圆润而轻盈

shifting—changing from one thing to another

千姿百态——不断变化成各种形状

towering—rising to a tall height

高耸——升到高处

vapor—tiny droplets of water floating in the air

水汽——轻飘在空中的小水滴

读书活动指导

1. 到户外看看。天空中有什么样的云呢？今天的天气如何？

2. 下次出门时，请抬头看看。你能看到云朵的形状吗？都有什么形状呢？

3. 请试着画一幅户外的风景画，画一朵这本书里介绍的云。